Drought and the Earth

Nikki Bundey

A ZOË BOOK

A ZOË BOOK

© 2001 Zoë Books Limited

Devised and produced by
Zoë Books Limited
15 Worthy Lane
Winchester
Hampshire SO23 7AB
England

Apart from any fair dealing for the purposes of research or private study, or criticism or review, as permitted under the Copyright, Designs and Patents Act, 1988, this publication may only be reproduced, stored or transmitted, in any form or by any means, with the prior permission in writing of the publishers, or in the case of reprographic reproduction in accordance with the terms of licences issued by the Copyright Licensing Agency.

Any person who does any unauthorised act in relation to this publication may be liable to criminal prosecution and civil claims for damages.

First published in Great Britain in 2001 by
Zoë Books Limited
15 Worthy Lane
Winchester
Hampshire SO23 7AB

A record of the CIP data is available from the British Library.

ISBN 1 86173 032 2

Printed in Italy by Grafedit SpA
Editor: Kath Davies
Design: Sterling Associates
Illustrations: Artistic License/Tracy Fennell,
　Genny Haines, Janie Pirie
Production: Grahame Griffiths

Photographic acknowledgments
The publishers wish to acknowledge, with thanks, the following photographic sources:

Sarah Errington - title page / John Hatt 14 / Maurice Harvey 15b / Isabella Tree 23t / Nigel Smith 23b / Jesco von Puttkamer 27 / Hutchison Picture Library; Francesco Rizzoli - cover (inset) right / Alan Keohane 12 / Tim Fisher 15t / Marco Siqueira 17t / Mark Henley 17b, 29b / Yann Arthus Bertrand 21b / Javed A Jafferji 24 / Impact Photos; A.N.T. 13b / K Ghani 16 / Karl Switak 25t / NHPA; Klein/Hubert - cover (inset) left / Baker-Unep 4 / NASA 5t / Ron Gilings 5b / Fabrice Beauchene 7t / Mike Schroder 7b / Mark Edwards 8t / Theresa de Salis 9 / Philippe Racamier 10 / Nigel Dickinson 18t / Tony Crocetta 18b / M & C Denis-Huot 19 / Gil Moti 22 / Cyril Ruoso 25b / Voltchev-Unep 28 / Manfred Gorgus 29t / Still Pictures; I Hoath - cover (background) / Michael Taylor 8b / B Turner 13t / J Sweeney 21t / TRIP.

The publishers have made every effort to trace the copyright holders, but if they have inadvertently overlooked any, they will be pleased to make the necessary arrangement at the first opportunity.

CONTENTS

Here comes the Sun	4
Sun and Earth	6
Rays of light	8
How hot is it?	10
Warming the Earth	12
Heat and moisture	14
Drought!	16
Dry as dust	18
In the desert	20
Plants in dry lands	22
Animal survival	24
Greenhouse gases	26
Turn down the heat	28
Words we use	30
Index	32

All the words that appear in **bold** type are explained in Words we use on page 30.

HERE COMES THE SUN

The Sun rises in the east, at dawn. Sometimes we see it in a clear sky. Sometimes **clouds** hide the Sun, but it still lights up the sky. The Sun reaches its highest point at midday, and then begins to sink. At sunset, it disappears from view, in the west.

The Sun brings light and warmth to Earth, our **planet**. Light and heat are types of **radiation**. Without them, no living thing could survive on Earth.

We all welcome a warm, sunny day. But if the Sun is too hot, it can make life very uncomfortable.

Renew online at http://www.catalogue.sefton.gov.uk/.

Or by telephone at any Sefton library:
Bootle: 0151 934 5781   Meadows: 0151 288 6727

Crosby: 0151 257 6400   Netherton: 0151 525 0607

Formby: 01704 874177   Southport: 0151 934 2118

A fine will be charged on any overdue book plus the cost of reminders sent

The Sun is a gigantic ball of burning hot **gases**, spinning around in space. It has a **mass** 330,000 times the size of our planet. If the Sun can burn us here on Earth, imagine how hot it must be at its surface.

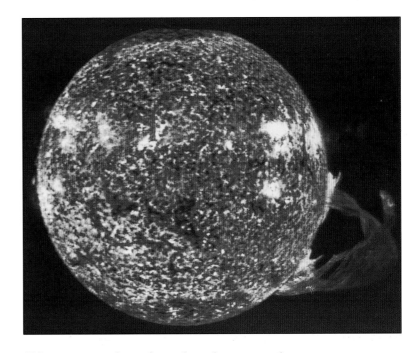

Plants and animals also need water to stay alive. The Sun's heat can dry up rivers and lakes. A long period of dry weather is called a **drought**.

The Sun is Earth's nearest **star**. It is about 150 million kilometres out in space, but it causes the weather conditions here on Earth. The Sun even helps to make the rain which ends a drought.

The Sahel region of Africa lies to the south of the world's biggest desert, the Sahara. Droughts here can last for years on end. Without enough water, crops and cattle die. People may starve.

SUN AND EARTH

The Earth moves around the Sun. Its path, or orbit, takes about a year to complete. As it journeys through space, the Earth also spins round on itself, once every day.

The earth tilts as it spins. As parts of the earth tilt towards the sun, they have the warmth of the summer. As they tilt away from the sun, they experience the cold of winter. Winter, spring, summer, and autumn are called **seasons**.

The **rays** of the Sun strike the middle of the Earth, around the **Equator**, directly. Here, the weather is mostly very hot. The Sun's rays strike the **Poles** at a slant or angle. The rays pass through a broad layer of the **atmosphere**. Here, the weather is bitterly cold.

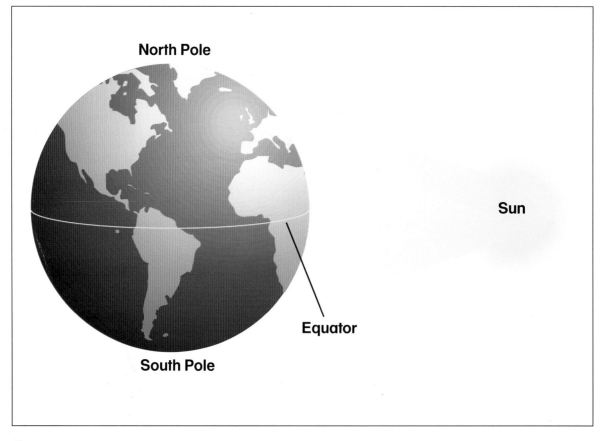

Polar regions receive very little warm sunshine. There is not much liquid water for living things except during the short summer. The water stays frozen as **solid** ice.

A layer of gases, called the atmosphere surrounds the Earth. This is the air we breathe. The atmosphere is often dusty and the dust blocks out some of the Sun's rays. Clouds reflect some rays away from Earth, too.

Without this protection, the surface of the Earth would be baking hot. As it is, only half of the Sun's radiation reaches the Earth.

Drought conditions are normal in some hot and dry regions, such as Monument Valley Desert in Arizona, USA.

RAYS OF LIGHT

The Sun sends out rays of light. We see this type of radiation every day. Daylight comes when the part of the Earth where we live spins round to face the Sun.

Light takes eight minutes to reach us from the Sun. It always travels at the same speed, 300,000 kilometres per second. If the Moon passes between the Earth and the Sun it blocks out the light. This is called an **eclipse**.

The rays of the Sun are so bright that they can damage your eyes, even during an eclipse. These people are wearing special glasses to protect their sight as they watch an eclipse.

A ray of light usually travels in a straight line. However it can be bent. This happens when sunlight passes through a raindrop. The light splits into seven different colours, forming a rainbow.

Sunlight makes it possible for plants to grow. Their leaves contain a green chemical called **chlorophyll**. It works with sunlight to make food for the plant.

Plants drink up water from the soil and pass it out into the atmosphere as an invisible gas called **water vapour**. By helping plants grow, sunlight plays an important part in creating moist air conditions.

Without sunlight, there would be no green plants on Earth. Without plants to eat, there would be no animals or people. The surface of Earth would be a dry, dusty desert, like the surface of the planet Mars.

See for Yourself

- Place a plant in a pot on a sunny window ledge.

- Over the next few weeks, does the plant lean towards the sunlight?
- Do leaves grow better on the sunny side or the shady side of the plant?
- What do you think might happen if you put the plant in a dark cupboard?

HOW HOT IS IT?

The Sun warms land and sea. We measure their heat and call this their **temperature**. Heat is measured in **degrees** (°) on a **scale** such as Celsius (C) or Fahrenheit (F). The **freezing point** of water is 0°C or 32°F.

We test weather conditions by measuring air temperature. The highest air temperature, 58°C, was recorded at El Aisisa in Libya. The hottest average temperature for a year was 34°C, recorded at Dallol in Ethiopia.

We can feel the heat of the Sun with our bodies, but we cannot usually see it. Sometimes heat affects gases in the atmosphere and makes the air shimmer.

We call a long period of hot, dry weather a heat wave. The pattern of weather recorded in one place over a long time is its **climate**. Regions with a dry climate, such as deserts, are called **arid zones**. Droughts often happen where the weather is hot and dry.

We use different types of **thermometer** to measure temperature. Most are glass tubes containing a **liquid** such as alcohol. As the alcohol becomes warmer, it gets bigger, or **expands**, and is forced up the tube. As it cools, it shrinks, or **contracts**, and drops back down.

See for Yourself

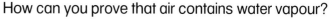

How can you prove that air contains water vapour?
- Place a household thermometer outside in a shady place.

- Record the air temperature at the same time each day during the summer months.
- Which was the highest temperature you noted down?
- Which was the lowest temperature you noted down?
- Which month was the hottest?
- Which month was the coolest?

June: Temperature at 8a.m.											
1	20°C	6		11		16		21		26	
2	21°C	7		12		17		22		27	
3	19°C	8		13		18		23		28	
4	21°C	9		14		19		24		29	
5	24°C	10		15		20		25		30	

WARMING THE EARTH

The land takes in, or **absorbs**, heat from the Sun. Rocks and soil do not carry or **conduct** heat easily. This means that usually, only the upper layers of the Earth's surface are warmed by the Sun.

The surface heat is mostly gentle but, under a blazing sun, desert rocks and sand can reach as much as 82°C. That is hot enough to fry an egg. As the ground cools later on, the heat radiates back into the air.

Rocks can reach very high temperatures by day. At night they quickly lose their heat through radiation.

The Sun heats the oceans to a depth of several hundred metres. Water holds heat well, so the oceans help to keep the world warm. Hot water from areas near the Equator changes place with cold water from polar regions. These movements cause ocean currents.

The wind is racing across this landscape in Australia. It is whipping up the dust into a column called a dust devil.

When the Sun heats the gases in the atmosphere, they rise. Cooler **air currents**, or winds, then rush in below the rising air. Winds often dry out moist land, turning the soil into dust. Dust may hang in the air and be blown for hundreds of kilometres. Wind makes periods of drought even harder to bear.

HEAT AND MOISTURE

Heat from the Sun dries up large amounts of the water in seas, lakes, rivers and puddles. The water doesn't disappear. It turns into the gas called water vapour. We say it **evaporates**.

The Sun's warmth makes the water vapour rise. The gas cools and now turns back, or **condenses**, into liquid water. The tiny water **droplets** hang in the air as mist or clouds, before falling to Earth as rain. Rain is the chief source of all the water on Earth.

This is Lake Assal in the hot, dry land of Djibouti. The fierce heat of the Sun makes the lake water evaporate. When the liquid turns to vapour, it leaves behind minerals and salts that were in the water.

The pattern of winds across the Indian Ocean gives Southeast Asia a warm, humid climate. Lush **rainforests** grow here and drought is unknown.

When the air contains large amounts of water vapour we say it is **humid**. In **tropical** lands, the air holds a lot of moisture from the ocean. These areas can be very hot and humid. Land in the centre of big **continents**, far from the sea, can be either hot or cold, but it is nearly always dry.

The tropics always receive direct sunlight, so there is no winter or summer there, only a dry season and a wet season. The dry season leaves tropical East Africa's rivers empty. The wet season comes when winds carry moisture from the ocean and drop it as rain on the land.

DROUGHT!

A drought is a long period with little or no rain. Rivers and lakes dry up, because there is no moisture left in the ground.

In tropical countries which have dry and wet seasons, the rains do not always arrive. Arica, in Chile, recorded no **rainfall** at all for 15 years. Even in countries which normally have mild and moist climates there may be droughts, especially during the summer months.

Many rivers in India hold only a trickle of water during the dry season. They may dry up completely during a drought. The winds of the wet season are called **monsoons**. They drop rain from the Indian Ocean, and may cause rivers to flood.

Life without water is impossible. Lakes and **reservoirs** dry up during drought. The fish died in this reservoir in Brazil when the water dried up. A long drought may kill plants, wildlife, cattle and even people.

Large areas of high **air pressure** are called anticyclones. They bring fine, dry weather. Rain clouds form in areas of low air pressure, called depressions. These **weather systems** whirl around the world, one taking the place of the other.

A powerful anticyclone may stop a depression from moving in. Great air masses build up over continents. They may cover an area of a million square kilometres and not move. The months pass by, and the rains never come.

Droughts do not always happen during periods of hot weather. The deserts of Central Asia can be bitterly cold and dry. A woman and her children carry water home during a winter drought in Mongolia.

DRY AS DUST

In times of drought, moisture evaporates from the soil. The soil cannot stick together, so it cracks and crumbles. The hot sun bakes the mud as hard as concrete. When the rains do come, the water cannot soak into the ground easily. It runs away over the hard surface and may cause sudden, flash floods.

Drought conditions can change a landscape. If the drought goes on and on, the land may turn into desert.

This was once the floor of a reservoir. During a long drought, the water has disappeared. The Sun, has baked the mud and caused patterns of deep cracks.

Dust fills the air around this village in Mali, Africa. During a drought, the wind can blow the dust over long distances. When it settles, a layer of dust may cover houses and fields.

These rocks are part of the Ahaggar Mountains in Algeria. Sand and grit in the wind has worn the rock into these towering pinnacles.

If plants die in a drought, their roots no longer hold the soil together. The wind may blow the soil away as dust and grit, leaving only bare rock.

The wind blows the grit hard against the surface of other rocks. This rubbing force is called **friction**. It can wear down, or **erode**, the rocks into strange shapes.

See for Yourself

- Mix soil with water in a flower pot, to make a 'mud pie'.
- Leave it in a hot, dry place.
- How long does it take to dry out?
- What has happened to the water the pie contained?
- Does it look different? How?

IN THE DESERT

Deserts are lands where there is drought year after year. Almost one-third of the Earth's surface is taken up by arid zones. Here, less than 300 millimetres of rain falls in a year.

Many deserts lie in the tropical lands to the north and south of the Equator. These are areas of high air pressure. They include the Sahara and Kalahari Deserts of Africa and the deserts of Arabia.

Arid zones called **rain-shadow** deserts form in some regions, such as the western United States.

The Taklamakan Desert in western China is cold in winter and very hot in summer. It is far from the ocean, so there is little water vapour in the air. No blanket of cloud protects the land from the 3,000 hours of sunshine it receives each year.

The deserts of Central Asia are in the middle of the continent, where few ocean winds shed rain.

The Namib Desert of Africa and the Atacama Desert of South America are beside cold ocean currents. But the cool sea air does not rise to produce rain. Instead, it is drawn over continents where the Sun scorches the land. Heat radiating from the land soon evaporates the moisture in the air.

Mist and clouds form where the cold Atlantic Ocean air meets the heat of the Namib Desert. However, the rainfall is less than 12 millimetres a year.

PLANTS IN DRY LANDS

The type of plants growing in a region depends on the rainfall. Forests need high rainfall. Grasslands need less moisture. In very dry areas, only scrub or desert plants can grow.

Many plants have **adapted** to life in dry places. Small flowering plants grow in cracks in the rock, where **dew** gathers and there is some shade. Baobab trees grow on the African grasslands, where drought is common. They store water in huge, swollen trunks.

In a drought there is no water pushing through the plant's stem and leaves. It droops or wilts in these conditions. Without green chlorophyll to make food for plants, they turn brown and die.

Cactus plants can survive in extremely dry conditions. They store moisture in tough, fleshy stems. Their sharp spines protect them from thirsty desert animals.

Even in a drought, pools of water may lie deep underground. Many arid zone plants have long roots which reach down to this moisture. Mesquite roots grow down 20 metres or more.

Seeds cannot sprout, or **germinate**, in hard, dry soil. Some desert plants have a tough seed coating which protects them through years of drought. When the rain comes, it washes away the coating and the seeds grow.

Agave plants are arid zone plants with tough leaves. Their waxy surface prevents them from losing too much water through evaporation.

ANIMAL SURVIVAL

Drought can be a disaster for wildlife. Without water, plants and animals which live in lakes and rivers may die. The animals that feed on them will die too.

Some animals have adapted to heat and drought. The little fennec fox lives in North Africa. The fox loses heat from its body through its huge ears. Like many desert creatures, it burrows underground away from the Sun. The fox comes out at night, when it is cool.

Crocodiles are **cold-blooded** animals. They need the Sun's warmth to keep their bodies working at the right temperature. They lie, or bask, in the sun. If they get too hot, crocodiles open their jaws wide to let out some body warmth.

This side-winder snake lives in the deserts of the USA. It moves by looping its body sideways over the ground. In this way, it makes as little contact as possible with the burning hot sand.

Some animals slow down their bodies during long, dry seasons. Their bodies go into a state rather like sleep, called **aestivation**. Some fish and frogs can survive drought in this way. They bury themselves in the mud of dry river beds until the water returns.

The addax antelope lives in the Sahara Desert. The addax can go for long periods without drinking water. It takes in moisture from dew and from the plants it eats.

GREENHOUSE GASES

Greenhouses are buildings made of glass. They are designed to trap sunlight and warmth so that tender young plants can grow inside them. Scientists now believe that heat from the Sun is trapped around our planet in the same way. They call this the **greenhouse effect**.

People have poisoned, or **polluted**, the atmosphere. They have filled the air with exhaust fumes from cars and smoke from factories. They have burned down forests and used the land for farming and for cattle ranching.

Air pollution has increased the amount of certain gases around the Earth. These gases hold in the warmth around the planet, so our patterns of climate are changing.

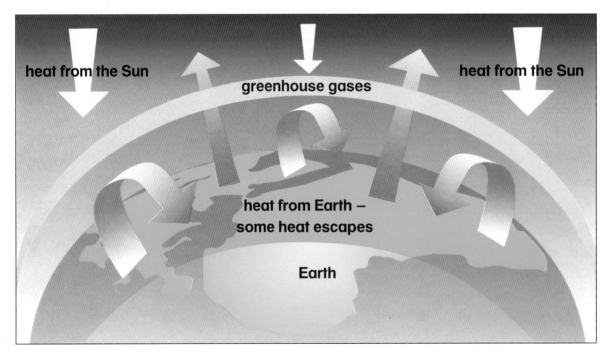

People have started fires in the rainforests of Brazil. Some areas are covered in **smog**. The atmosphere is full of gases such as carbon dioxide and nitrous oxide. These are called greenhouse gases.

These activities have increased the amount of gases such as methane, carbon dioxide and nitrous oxide in the atmosphere. The gases have formed a blanket which **insulates** the planet. They let through some of the Sun's rays, but they also stop heat from escaping. The gases reflect heat back to the Earth's surface.

See for Yourself

- Make a list of things that pollute the air in your neighbourhood. Examples might include
  - traffic jams
  - factory chimneys
  - garden bonfires
- Can you think of any ways of reducing local pollution?

TURN DOWN THE HEAT

The Earth's climate is always changing. In the past there have been long **Ice Ages**, followed by warmer periods. Sometimes weather conditions have changed because a volcano has erupted and filled the atmosphere with ash.

Today the climate is changing very quickly. The greenhouse effect is making our planet overheat. This overheating is called **global warming**.

Around 7,000 years ago, the Sahara region of Africa was lush and green. Today it is the world's biggest desert. We must learn to understand more about the way in which Earth's climate can change. Our future depends on it.

This land in Spain is turning into desert. Large areas of southern Europe, where the present climate is warm but mild, may become arid zones.

As the world heats up, droughts will become more common. Deserts will spread in many parts of the world. At the same time, other parts of the world will become wetter and more stormy. Polar ice will melt and the sea water level will rise. Sea water will flood over low land. Many types of plant and animal could become **extinct**.

Global warming is already beginning to happen. It can be reduced only if people are prepared to stop polluting the atmosphere.

29

WORDS WE USE

absorb	Take in or soak up.
adapt	To change in order to survive in particular conditions.
aestivation	A state of drowsiness that helps animals survive during a drought.
air current	A movement of air, wind.
air mass	A large amount of air which does not move.
air pressure	The force with which air presses down on the planet's surface.
arid zone	One of the world's very dry regions.
atmosphere	The layer of gases around a planet.
chlorophyll	A green substance in plants which uses sunlight to help them make food.
climate	The pattern of weather in one place over a long period.
cloud	A mass of water droplets or crystals hanging in the air.
cold-blooded	Having a body which cannot make its own warmth.
condense	To turn from gas into liquid.
conduct	To carry or pass on heat or electricity.
continent	A large mass of land, such as North America or Africa.
contract	To take up less space, to shrink.
degree	A unit on a scale, such as those used to record temperatures.
dew	Tiny drops created when water vapour condenses on cold surfaces.
droplet	A tiny drop water. Many droplets make up a raindrop.
drought	A long, dry period with little or no rainfall.
eclipse	When the Moon moves between the Sun and the Earth and blocks out sunlight, or when the Earth's shadow blocks reflected light from the Moon.
Equator	An imaginary line which mapmakers draw around the middle of the Earth.
erode	To wear down rock or soil by wind, water, frost or ice.
evaporate	To turn from a liquid into a gas.
expand	To take up more space, to swell.
extinct	No longer in existence, having died out.
freezing point	The temperature at which water turns to ice (0°C).

friction	The force which slows one object down as it rubs against another.
gas	Any airy substance which fills any space in which it is contained.
germinate	To begin the growth from seed to plant.
global warming	The gradual heating up of our planet.
greenhouse effect	Trapping warmth around the Earth, in the same way as a greenhouse traps heat from the Sun.
humid	Containing a lot of water vapour.
Ice Age	A time of global cooling, when ice spreads far from the polar regions.
insulate	To use a cover to prevent heat loss, or protect from electricity or sound.
liquid	A fluid substance, such as water.
mass	The amount of matter in an object.
monsoons	Seasonal rain-bearing winds in southern Asia.
planet	One of the worlds travelling around the Sun, such as Earth.
polar	To do with the most northerly and southerly parts of a planet, around the Poles.
Poles	The most northerly and southerly points on a planet.
pollute	To poison land, water or air.
radiation	Giving out rays, such as heat or light.
rainfall	The amount of rain measured in a particular place over time.
rainforest	An area of forest in a region of high rainfall.
rain-shadow	The dry region behind a coastal mountain range which receives very little rainfall.
ray	A beam of radiation, such as light.
reservoir	A lake or large tank used to store water.
seasons	Climate variations caused as the Earth tilts and travels around the Sun. Spring, summer, autumn and winter are seasons.
scale	A series of graded units.
smog	A mixture of fog and smoke or exhaust fumes.
solid	Something which is hard or firm, and has length, height and width.
star	A glowing ball of gas in space, which radiates heat and light.
temperature	Warmth or coldness, measured in degrees.
thermometer	Any instrument used to measure temperature.
tropical	To do with the Tropics, the regions to the north and south of the Equator.
water vapour	A gas created when water evaporates.
weather system	A set of weather conditions, such as an anticyclone or depression.

INDEX

aestivation, 25
air currents, 13
air pressure, 17, 20
animals, 5, 9, 17, 23, 24, 25
anticyclones, 17
arid zones, 11, 20, 22, 23
atmosphere, 6, 7, 9, 10, 13, 26, 27, 28

chlorophyll, 9, 22
climate, 11, 15, 26, 28, 29
clouds, 4, 7, 21
condensation, 14
continents, 15, 17, 21

depressions, 17
deserts, 5, 7, 9, 12, 17, 18, 20, 21, 22, 23, 25, 28, 29
dew, 22, 25
drought, 5, 7, 11, 13, 15, 16, 17, 18, 22, 23, 24, 29
dry seasons, 15, 16, 25
dust, 7, 13, 18, 19

Earth, 4, 5, 6, 7, 8, 9, 12, 14, 26, 27, 28
eclipses, 8
Equator, 6, 20
erosion, 19
evaporation, 14, 18, 21, 23

flash floods, 18
friction, 19

gases, 5, 7, 9, 10, 13, 14, 26, 27
global warming, 28, 29
greenhouse effect, 26, 27, 28

heat waves, 11
humidity, 15

light, 4, 8, 9

mist, 21
monsoons, 16

ocean currents, 13, 21
oceans, 13, 15, 16

plants, 5, 9, 17, 22, 23, 24, 25, 26
Polar regions, 6, 7, 29
pollution, 26, 27, 29

radiation, 4, 8, 12, 21
rain, 14, 15, 16, 17, 20, 21, 22, 23
rainbows, 8
rainforests, 15, 27
rays, 6, 7, 8, 27
reservoirs, 17, 18
rock, 12, 19
roots, 19, 23

seasons, 6, 7, 15, 16, 21
seeds, 23
smog, 27
soil, 9, 12, 13, 18, 19, 23
Sun, 4, 5, 6, 7, 8, 10, 12, 13, 14, 18, 26, 27

temperature, 10, 11, 12, 24
thermometers, 11
tropical lands, 15, 16, 20

water vapour, 9, 11, 14, 15, 21
weather systems, 17
wet seasons, 15, 16
winds, 13, 15, 18, 19, 20